HOW IT HAPPENS

at the Cereal Company

By Megan Rocker
Photographs by Bob and Diane Wolfe

Minneapolis

County Market[SM]

Dear Reader,

It takes a tremendous amount of work – and a lot of dedicated workers – to make delicious and nutritional Malt-O-Meal cereals. This book highlights the different steps involved in making Malt-O-Meal cereal – before it arrives on the shelves at County Market stores. Malt-O-Meal takes great care to manufacture a variety of top-quality cereals and County Market makes them available at bottom-line prices – a winning combination!

We hope you'll enjoy your book tour through the Malt-O-Meal plant. Thanks for taking an interest in Malt-O-Meal Cereals and look for these fine cereals at your local County Market store.

Sincerely,

County Market and Malt-O-Meal

Cereal is many people's favorite breakfast food. It's healthy and easy to prepare, and it comes in enough varieties to please almost everyone. Some cereal eaters prefer simple corn or wheat flakes, others crave kinds with nuts, raisins, or berries, and still others love frosted, brightly colored cereals in fun shapes.

Cereal can taste like chocolate, cinnamon, or a rainbow of fruit flavors. It can be crunchy, puffy, or filled with marshmallows. How do cereal companies create so many different, delicious products? This book will show you step by step how a package of breakfast cereal reaches your table.

Ingredients

Making cereal starts with basic ingredients. Some are familiar ones you might find in your kitchen, such as flour, oil, sugar, and salt. Others are special ones you may not have seen before. For example, **vitamins and minerals** (such as iron) make the cereal more nutritious, and dyes make it more colorful. This company makes many kinds of cereal, so it uses a variety of ingredients to create different looks, tastes, and textures.

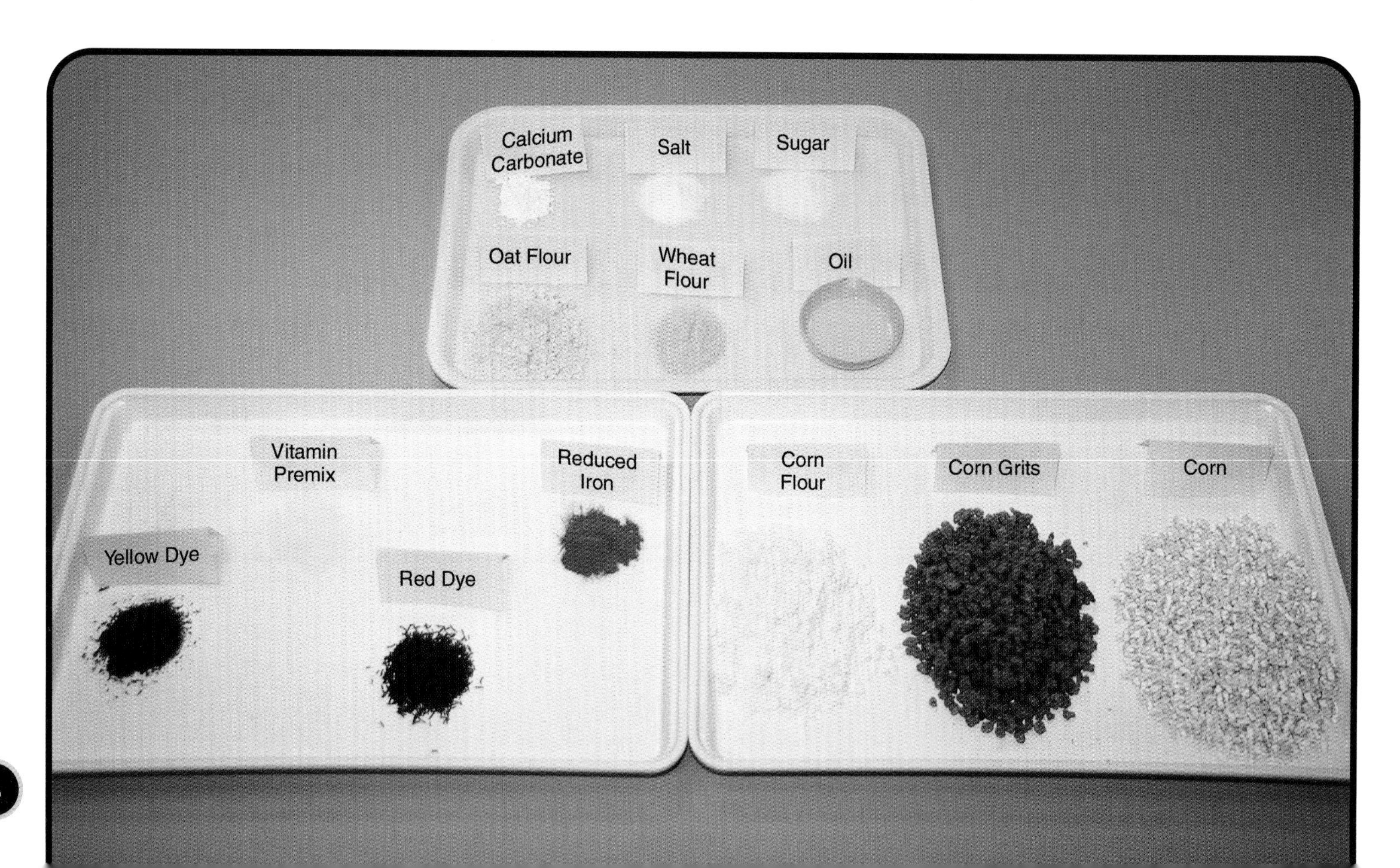

Many ingredients, including flour and sugar, come to the cereal factory in large trucks. Long hoses carry the ingredients out of the trucks and into storage bins.

These huge tanks hold sweeteners like molasses and corn syrup until they are ready to be used.

Workers use computers to control the machines that move ingredients from place to place and mix them together to make cereal. The computer shown here runs the cooking machines.

Cooking Corn

Some cereals are made directly from cooked **grains** such as corn and rice. This picture shows raw corn being poured into a large vat called a cooker.

The lid of the cooker is sealed, and the corn is cooked with steam until it softens. The cooker turns slowly so the corn is heated evenly.

Making Flakes

Next, the cooked corn is stirred to break up the large lumps into smaller clumps, called **grits**. Once they are dried, the grits will be made into corn flakes.

A machine called a mill presses the grits flat between two large metal rollers.

Next, the flakes pass under hot jets of air that bake them until they are toasted and crispy.

Extrusion

Other cereals are made by a process called **extrusion**. First, ingredients such as flour, sugar, salt, and vitamins are poured into a machine.

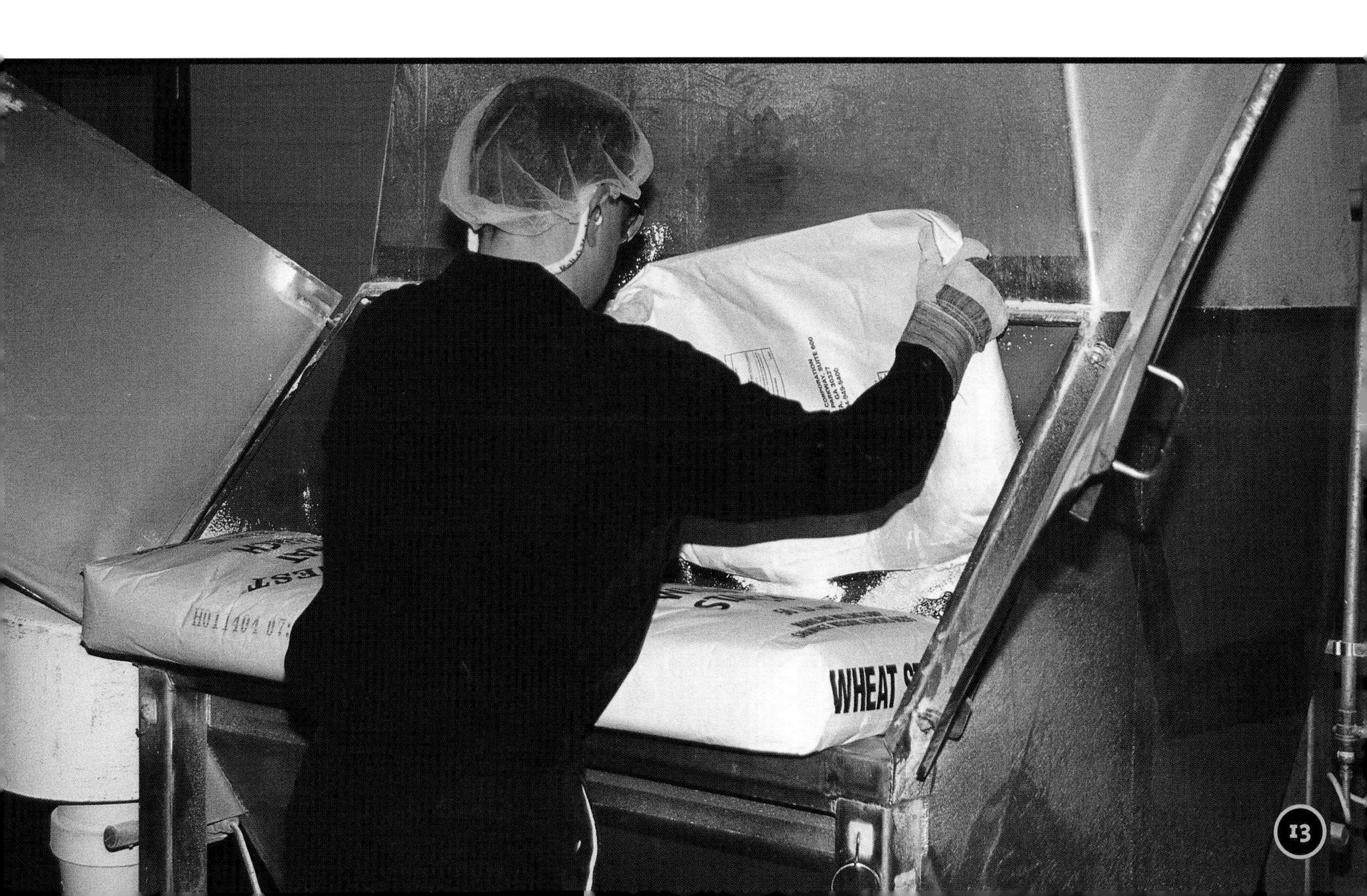

The machine mixes the ingredients together and cooks them to make a dough.

Next, the dough (left) is formed into shapes. To make an O-shaped cereal, a machine called a forming extruder presses the dough through small O-shaped holes. The dough comes out in long tubes, which are then sliced into many individual pieces (right). The cereal is called **half-product** at this stage because it is not yet finished.

Coloring

To make colored cereals, dye must be added to the dough before it is shaped. A worker carefully measures the dye powder and mixes it with water.

Food dye has no taste. Dough colored with the red dye shown here will not have a different flavor than dough colored with green dye. The flavors of colored cereals are added later in the process.

Half-product is much smaller and denser than the finished cereal will be. It needs to be exposed to high temperature and pressure to make it puff up. Once this is done, the cereal may be given a flavored coating. This tray shows the dramatic difference between half-product (left), puffed but uncoated cereal (center), and the finished product (right).

Flavoring

Colored cereals get their fruit flavor from a liquid coating that is sprayed onto them inside a rotating drum. Drums like this one are also used to add the sweet coating to cereals such as frosted corn flakes, or to add more vitamins.

This cereal is finished now, but some other kinds of cereal have extra ingredients.

A machine like this one mixes in add-ons such as marshmallows (shown here) or raisins to complete the cereals that need them.

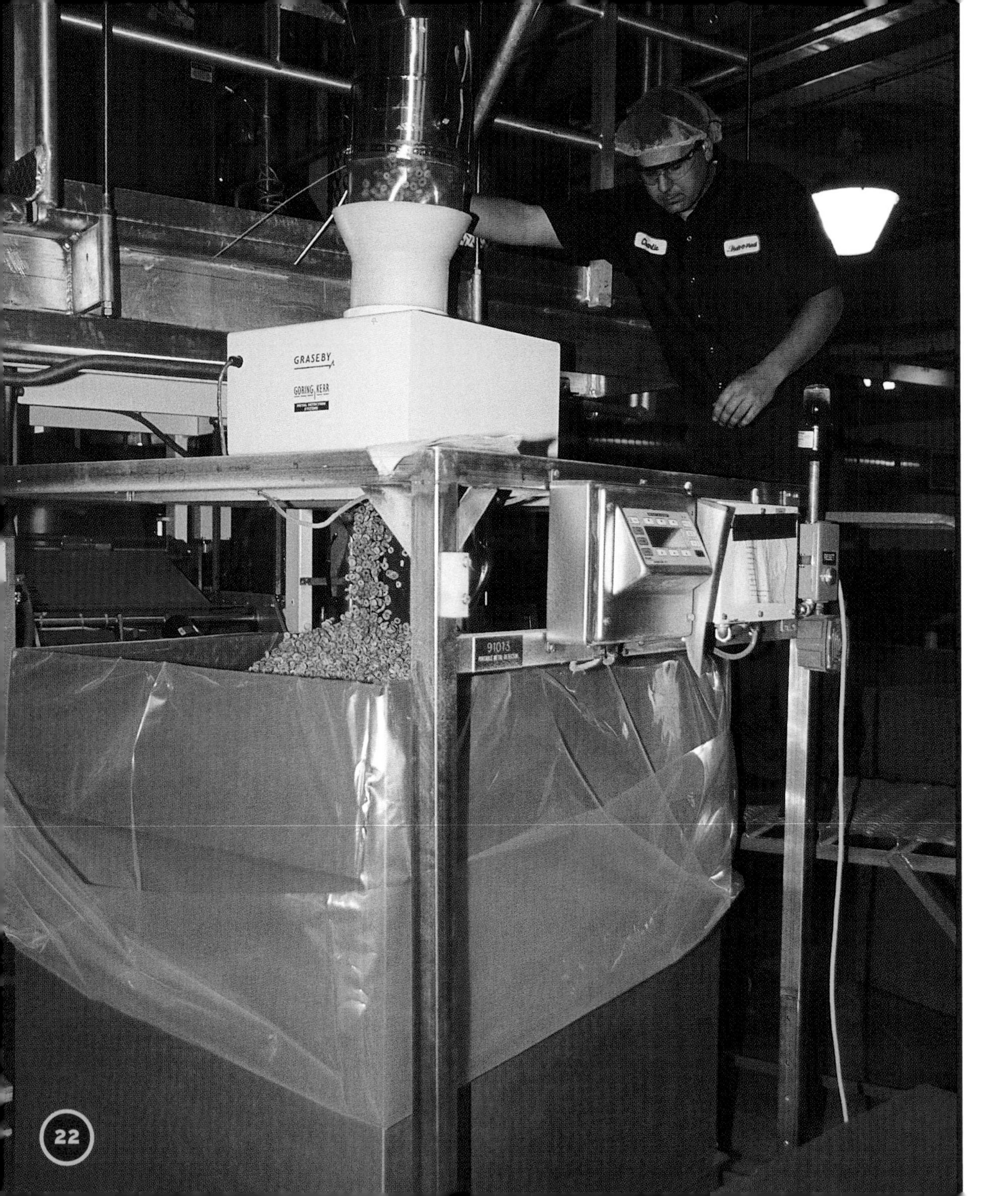

Packaging

The finished cereal is carried through pipes to the packaging areas. Some of it is poured into huge boxes to be used for filling single-serving-sized containers.

Most of the cereal is packaged in bags. The plastic for the bags comes in large rolls that are unwound by a machine.

The machine folds the plastic into a tube, drops cereal into it, and seals it.

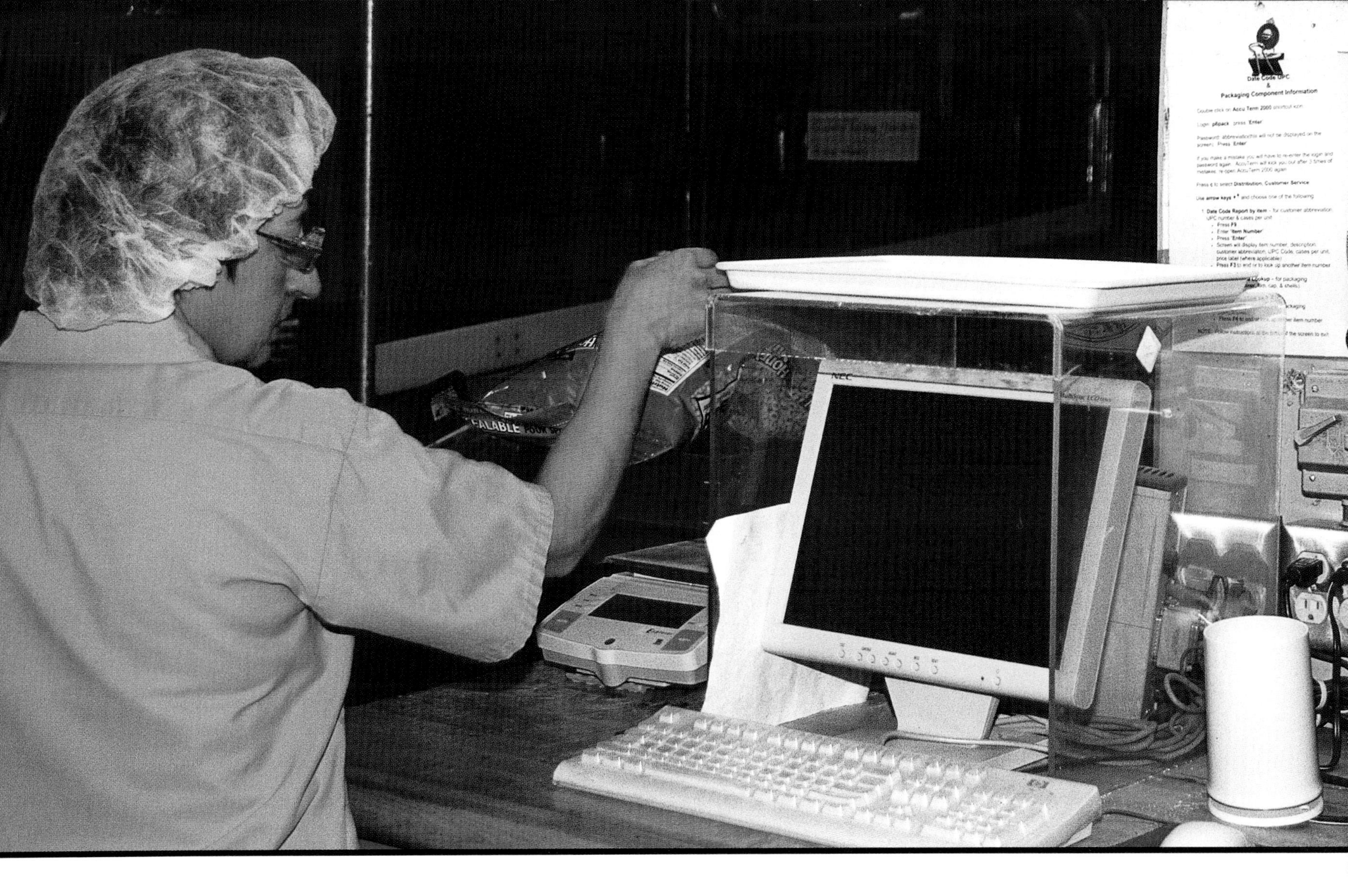

A worker randomly chooses bags of cereal and weighs them to make sure they are properly filled. She also tests to see if the bag is sealed securely. This process is called **quality control**.

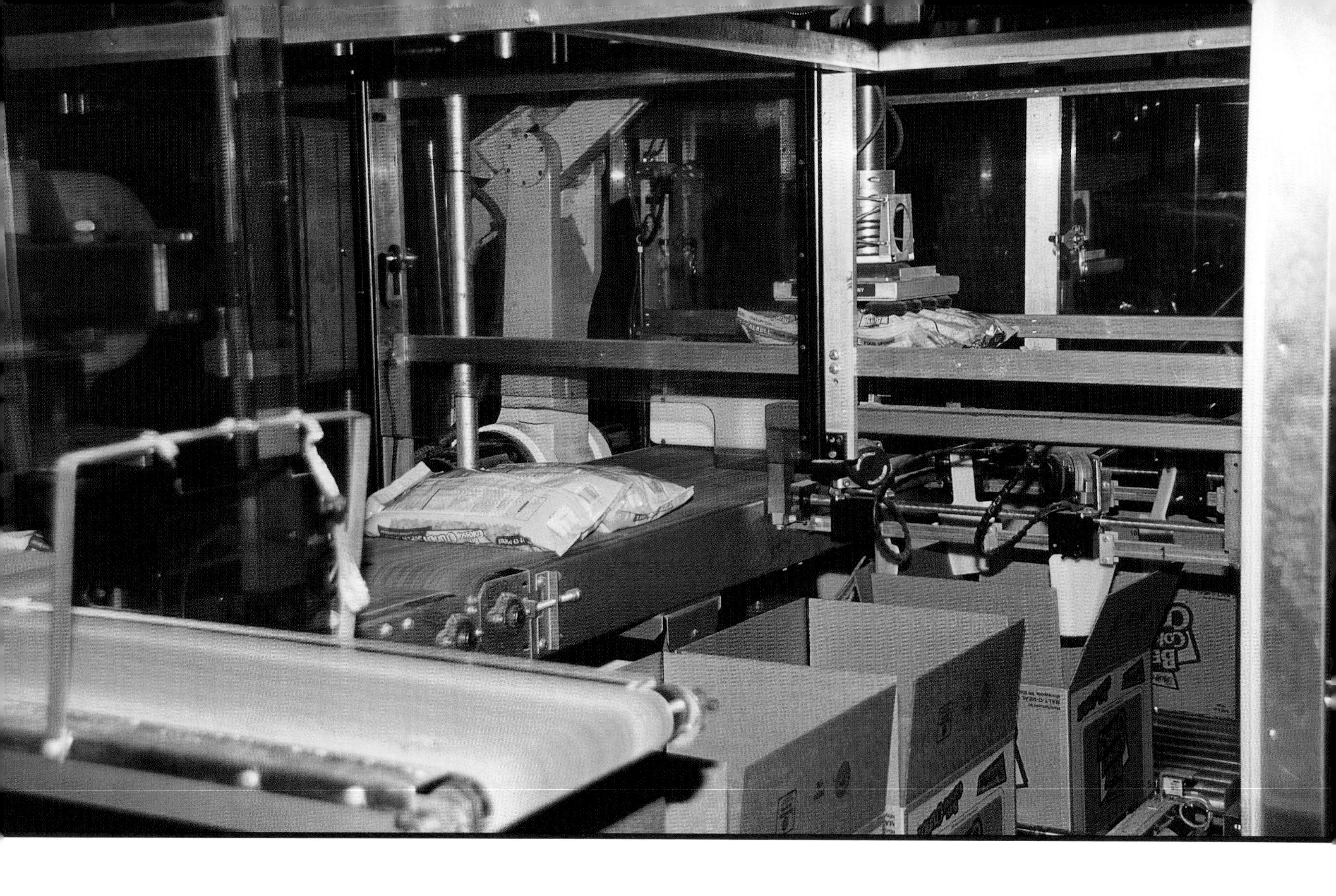

Packaging is a repetitive task that requires heavy lifting, so much of the work is done by machines called **robots**. Here, a robot packs bags of cereal into boxes to protect them during shipping.

Another robot closes the boxes and tapes them shut (top). Then a code is printed on the side of each box that includes the date and time it was packed (bottom). This helps the company give its customers the freshest product possible.

A robot picks up the boxes and stacks them in a large pile.

The stacks of boxes are wrapped in plastic to keep them together. This makes them easier to move from place to place.

The boxes are kept in a **warehouse**, or large storage area, until they are ready to be shipped. Then a worker uses a **forklift** (a machine that lifts and carries heavy objects) to load the boxes onto trucks that will take them to grocery stores around the country.

Malt-o-Meal
Chocolate
CALCIUM
Malt-o-Meal
Maple & Brown Sugar
CALCIUM
Malt-o-Meal
Malt-o-Meal
Original
FOLIC ACID
Malt-o-Meal
PERFECT BALANCE
A Great Start to Every Day!

RESEALABLE
Raisin Bran
Raisin Bran
50% MORE
Frosted Mini Spooners
Frosted Mini Spooners
50% MORE
Frosted Mini Spooners
Frosted Mini Spooners
50% MORE
50% MORE

RESEALABLE
Malt-o-Meal
Toasted Cinnamon Twists
Toasted Cinnamon Twists
Honey Graham Squares
Apple Cinnamon Toasty O's
Marshmallow Mateys
50% MORE

RESEALABLE
Toasted Cinnamon Twists
Toasted Cinnamon Twists
Toasted Cinnamon Twists
50% MORE
Honey Graham Squares
Honey Graham Squares
Honey Graham Squares
50% MORE

Balance with Berries
Balance with Berries
Balance with Berries
REAL STRAWBERRIES
ORIGINALS

RESEALABLE
Honey Nut Toasty O's
Honey Nut Toasty O's
Fruity Dyno-Bites
Honey Buzzers
Coco Roos
50% MORE

Corn Bursts
Corn Bursts
Corn Bursts
50% MORE
Honey Graham Squares
Honey Graham Squares
Honey Graham Squares
50% MORE
Balance with Berries
Balance with Berries
Balance with Berries
REAL STRAWBERRIES
ORIGINALS

Corn Bursts
Golden Puffs
Tootie Fruities
Cocoa Dyno-Bites
Cocoa Dyno-Bites
50% MORE

RESEALABLE
Marshmallow Mateys
Marshmallow Mateys
Marshmallow Mateys
50% MORE
Berry Colossal Crunch
Berry Colossal Crunch
Berry Colossal Crunch
50% MORE
Tootie Fruities
Tootie Fruities
Tootie Fruities
50% MORE

RESEALABLE
Apple Zings
Berry Colossal Crunch
Crispy Rice
Puffed Rice
Raisin Bran
50% MORE

Marshmallow Mateys
Marshmallow Mateys
Marshmallow Mateys
50% MORE
Berry Colossal Crunch
Berry Colossal Crunch
Berry Colossal Crunch
50% MORE
Tootie Fruities
Tootie Fruities
Tootie Fruities
50% MORE

Glossary

extrusion: shaping a soft material by pushing it through an opening that has the desired shape

forklift: a machine that lifts and carries heavy objects from place to place

grain: the small, dry, starchy fruits harvested from grasses such as wheat, corn, rice, and oats. Cereal may be made from any one of these grains, or from a combination of several.

grits: small clumps of cooked corn

half-product: cereal that has been cooked and shaped but not yet puffed. Half-product is smaller and denser than the finished cereal.

quality control: a system of inspecting products to make sure they meet certain standards

robot: a machine that can be programmed to perform tasks

vitamins and minerals: nutrients needed to keep the body healthy. They are found naturally in some foods and are added to others during manufacturing. Common vitamins in cereal include A, C, D, riboflavin, niacin, and thiamin. Common minerals in cereal include calcium, iron, and zinc.

warehouse: a large storage area